AF449058

A special thank you to my friends Ashley and Staci who didn't think my request for pictures of animal poop (and animals that look like poop) was weird at all. Or if they did think it was weird, they didn't tell me.

And a huge thank you to my writing buddies, Carrie and Shae. You two really know how to polish a turd.

Photographs © spineback/depositphotos.com; okiepokey/depositphotos.com; harmonia101/depositphotos.com; lifeonwhite/depositphotos.com; JimCumming/depositphotos.com; michaklootwijk/depositphotos.com; farinosa/depositphotos.com; cavan/depositphotos.com; trubavin/depositphotos.com; casarda/depositphotos.com; Deerphoto/depositphotos.com; kaiskynet@gmail.com/depositphotos.com; brszattila@gmail.com/depositphotos.com; sserg_dibrova/depositphotos.com; fouroaks/depositphotos.com; schankz/depositphotos.com; civic_dm@hotmail.com/depositphotos.com; MazurTravel/depositphotos.com; sasimoto/depositphotos.com; tong2530/Adobe Stock; DadangKusuma/Adobe Stock; bennytrapp/Adobe Stock; NumediaPhoto/Adobe Stock; Mandy Creighton/Shutterstock.com; KanphotoSS/Shutterstock.com; Sainam51/Shutterstock.com; chinahbzyg/Shutterstock.com; Simon Shim/Shutterstock.com; Marut Khrueahong/Shutterstock.com; Katarina Christenson/Shutterstock.com; SergKul/Shutterstock.com; Cloebudgie/istockphoto.com; sbonk/istockphoto.com; izanbar/istockphoto.com; Bumphead Parrotfish defecating/David Fleetham/Alamy; Spraint van otter/Buiten-Beeld/Alamy; Steephead Parrotfish/David Fleetham/Alamy; White-tailed jackrabbit eating cecotrope/Rosanne Tackaberry/Alamy; wirestock/Envato.
Rabbit cecotrope photo © Staci Hill.

IS IT POOP?

Kizzi Roberts

M.S. Animal Science

LearningSpark EDUCATIONALPUBLISHING

Poop, poo, scat, feces, dung, excrement...whatever you call it, poop is important in the animal world.

All animals poop, but not all poop is the same. Some poop is big, some poop is small, and some poop doesn't look like poop at all. Some poop is useful, some poop is food, some poop is defensive, and some animals look like poop.

Animals that look like poop are trying to avoid predators. Most predators don't want to eat poop... but some do. Animals are interesting, and so is their poop.

As a pooper yourself, you might consider yourself a poop expert, but this book contains pictures of wild and weird poop as well as a few im-poo-sters.

So, step right up, but don't step in it.

It's time for the game with one question and one question only...

IS IT POOP?

NOT POOP!

The giant swallowtail caterpillar looks like bird poop to avoid predators. This is called *mimicry*. The caterpillar will eventually transform into a giant swallowtail butterfly. There are over 500 different kinds of swallowtail butterflies, and most start as caterpillars that look like poop.

It has a distinct smell, but...

IS IT POOP?

POOP!

Otter poop is called **spraint**. Otter spraint is easy to identify by the way it looks. It contains mostly fish bones and scales. Otter spraint can also be identified by its smell. It has a sweet, fresh fish smell.

IS IT POOP?

POOP!

Wombats are known for their unique poop. Their droppings are shaped like cubes.

Wombats have a special area in their intestines that compacts the poop into a cube. A single wombat deposits about 100 poop cubes every day.

It's round and brown, but...
IS IT POOP?

NOT POOP!

Owl pellets sometimes look like poop, but an owl **regurgitates** pellets through its beak. Pellets contain parts of an owl's food that it can't digest, including bones and fur. An owl usually produces one pellet per day.

It's a big mess, but...

IS IT POOP?

POOP!

Whales poop a lot! The largest whale in the world can produce over 50 gallons of poop during one bowel movement.

Whale poop is important to the ocean's **ecosystem**. Krill and plankton need whale poop for nutrients, especially iron. These tiny ocean creatures eat whale poop to live, then whales eat them. It's the circle of life!

This looks a little squishy, but...

IS IT POOP?

NOT POOP!

Arkys curtulus is the scientific name for the small bird dropping spider. This spider mimics bird poop to hide from **predators** and attract **prey,** like soldier flies. The spider keeps its legs tucked in to look like poop. When a fly gets close enough, the spider stretches out its front legs to grab the fly.

Life-size Arkys curtulus 

IS IT POOP?

NOT POOP!

Theloderma asperum is the scientific name for this frog, also known as: pied warty frog, hill garden bug-eyed frog, and bird poop frog.

This frog resembles bird poop as a way to **camouflage** itself from predators.

This looks nutty, but...

IS IT POOP?

POOP!

The Asian palm civet is a shy, **nocturnal** animal, also called a "weasel cat". It naturally eats coffee berries in the wild, then poops out whole coffee beans. These beans can be cleaned, dried, and roasted into coffee called kopi luwak.

Coffee berries

Civet poop

Kopi luwak

This looks a little shiny, but...

IS IT POOP?

POOP!

Bat poop is called guano. Guano is shiny due to all the insects bats eat.

Insects have shiny **exoskeletons.** Bats can't digest the exoskeleton. The shiny exoskeleton passes into their poop, giving the poop a sparkly appearance.

IS IT POOP?

Poop isn't the only thing I like. I eat anything that's wet and rotten.

POOP!

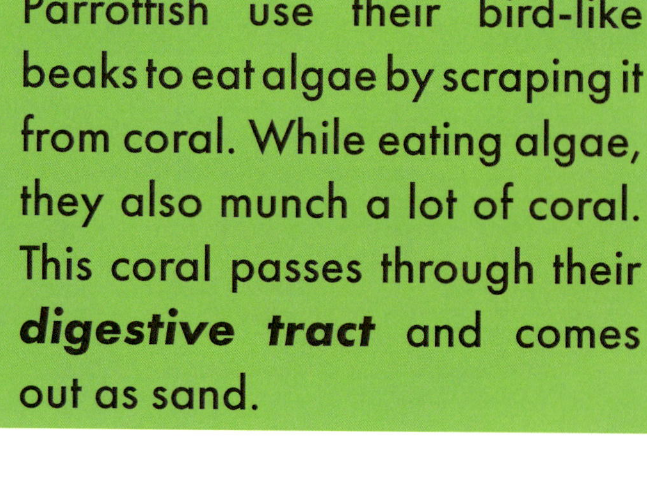

I chomp through coral at up to 20 bites per minute. All that coral makes me produce almost 200 pounds of sandy poop every year.

Parrotfish use their bird-like beaks to eat algae by scraping it from coral. While eating algae, they also munch a lot of coral. This coral passes through their **digestive tract** and comes out as sand.

This looks a little soft, but...

IS IT POOP?

That looks soft...
and delicious. Did you
know I can only
eat liquids?

NOT POOP!

Cecotropes are not poop! Rabbits make two types of pellets: fecal pellets, and cecal pellets (cecotropes).

Cecotropes are soft and contain food the rabbit was unable to fully digest. The rabbit eats cecotropes to get important nutrients. In fact, rabbits need cecotropes to live.

A rabbit often eats a cecotrope as it is expelled. It is rare to find a cecal pellet with the poop pellets.

This looks weird, but...

IS IT POOP?

POOP & NOT POOP!

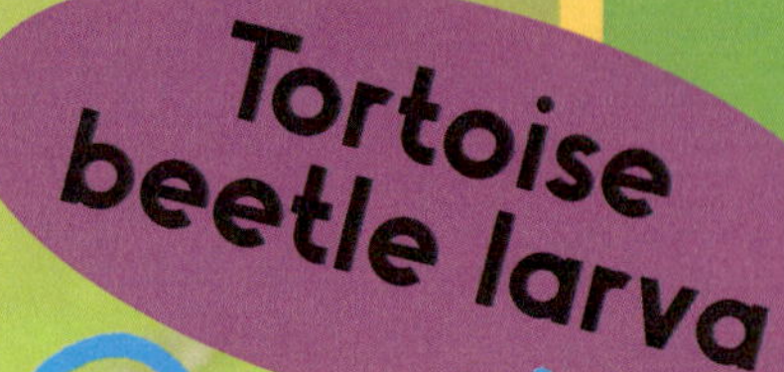

Poop parasol, feces pack, frass mask, fecal shield... there are many names for the unique defense mechanism used by tortoise beetle larvae.

There are many different kinds of tortoise beetle larvae, but they all make fecal shields.

Some larvae have two spikes at the tail end of their body. They use these tail spikes to carry shed skin and *frass* (poop). The shed skin and frass builds up on the tail spikes. This creates a fecal shield.

My fecal shield protects me from predators. It makes me look like poop.

Sometimes, I wave the shield around to act big and scary.

We've talked about poop, and things that look like poop, but what about me??? What about things that eat poop? Did you think about that?

House flies vomit and poop each time they land. They vomit saliva on solid food to turn it into a liquid. Flies can only eat liquid food. They eat poop and lay their eggs in it.

I'm a roller dung beetle. I roll dung into a ball and bury it to eat later. My larvae eat solid poop. I slurp up wet poop.

We are red wiggler worms. We eat poop and other organic waste. We turn the waste into worm castings, which are full of nutrients.

I'm a blind cavefish. Food sources are limited inside a cave, but bat poop is abundant. I have extra taste buds on my head and chin to help me find food.

GLOSSARY

camouflage	to hide by blending in with the surroundings or looking like something else
cecotrope	droppings filled with important nutrients
digestive tract	a winding tube that starts with the mouth and is used to break down food, absorb nutrients, and get rid of waste (poop)
ecosystem	the interaction of a group of organisms with each other and their environment
exoskeleton	a skeleton on the outside of an organism that provides support and protection
frass	insect poop
mimicry	when an organism looks like something else
nocturnal	active at night
predator	an animal that hunts and eats other animals
prey	an animal that is the food source for other animals
regurgitate	to bring swallowed food back up into the mouth
spraint	otter poop

AUTHOR'S NOTE

When I worked for the circus as a veterinary technician, I did a lot of interviews where I answered questions from kids about the circus animals. Something I learned is that kids are fascinated by poop. And I think they're on to something, because poop **IS FASCINATING** (and a little gross), but you know what else is fascinating (and a little gross)?

SCIENCE!

So, while this book is about poop, it's also about science... the science of mimicry and camouflage, and the science of nutrition and digestion. I earned my degrees in animal science in college where I learned about all sorts of different animals. I studied nutrition and digestive systems and everything else, but I didn't get the idea for this book until I had kids of my own.

One day I saw something on the floor. It was small and round and brown and I wondered...**IS IT POOP?**

It was actually a chocolate chip, but that made me think of everything I know about animals, everything I know about poop, and everything I know about how much kids love poop. And when I put all those ideas together, I came up with this book.

Thanks for reading, and I hope you enjoyed this book as much as I enjoyed creating it!

~ Kizzi

* 9 7 9 8 8 8 8 8 4 0 3 9 9 *